A BETTER APPROACH TO MOBILE DEVICES

How do we maximize resources, promote equity, and support instructional goals?

Susan
BROOKS-YOUNG

 Alexandria, VA USA

Website: www.ascd.org www.ascdarias.org
E-mail: books@ascd.org

Copyright © 2015 by ASCD. All rights reserved. It is illegal to reproduce copies of this work in print or electronic format (including reproductions displayed on a secure intranet or stored in a retrieval system or other electronic storage device from which copies can be made or displayed) without the prior written permission of the publisher. Readers who wish to duplicate material copyrighted by ASCD may do so for a small fee by contacting the Copyright Clearance Center (CCC), 222 Rosewood Dr., Danvers, MA 01923, USA (phone: 978-750-8400; fax: 978-646-8600; Web: www.copyright.com). To inquire about site licensing options or any other reuse, contact ASCD Permissions at www.ascd.org/permissions, permissions@ascd.org, or 703-575-5749. Send translation inquiries to translations@ascd.org.

Printed in the United States of America. ASCD publications present a variety of viewpoints. The views expressed or implied in this book should not be interpreted as official positions of the Association.

ASCD®, ASCD LEARN TEACH LEAD®, ASCD ARIAS™, and ANSWERS YOU NEED FROM VOICES YOU TRUST® are trademarks owned by ASCD and may not be used without permission. All other referenced trademarks are the property of their respective owners.

PAPERBACK ISBN: 978-1-4166-2164-5   ASCD product #SF116020

Also available as an e-book (see Books in Print for the ISBNs).

Library of Congress Cataloging-in-Publication Data
Names: Brooks-Young, Susan.
Title: A better approach to mobile devices : how do we maximize resources, promote equity, and support instructional goals? / Susan Brooks-Young.
Description: Alexandria, Virginia : ASCD, 2015. | Includes bibliographical references.
Identifiers: LCCN 2015033424 | ISBN 9781416621645 (pbk.)
Subjects: LCSH: Mobile communication systems in education. | Education--Effect of technological innovations on.
Classification: LCC LB1044.84 .B76 2015 | DDC 371.33--dc23 LC record available at http://lccn.loc.gov/2015033424

---

23 22 21 20 19 18 17 16 15   1 2 3 4 5 6 7 8 9 10

A BETTER APPROACH TO MOBILE DEVICES

How do we maximize resources, promote equity, and support instructional goals?

The Case for Hybrid Mobile Technology Initiatives 1

Planning for Success 11

Connection to Curriculum.. 14

Infrastructure and Support 17

Training and Professional Development.......................... 25

Budget 41

Policies and Procedures 47

Encore 53

References 58

Related Resources.. 61

About the Author 62

Want to earn a free ASCD Arias e-book?
Your opinion counts! Please take 2–3 minutes to give us your feedback on this publication. All survey respondents will be entered into a drawing to win an ASCD Arias e-book.

Please visit
www.ascd.org/ariasfeedback

Thank you!

The Case for Hybrid Mobile Technology Initiatives

Sharon Norman's high school biology students are completing a unit on homeostasis. She wants to know if they are able to apply what they've learned to real-world situations, so she devises an assessment activity that will take place over two periods. She breaks the class into teams of three, then randomly assigns each trio one of six examples of homeostasis in action (e.g., sweating, thirst). The students' assignment is to use one of eight iPads (checked out from the media center) to make a one-minute video explaining why their topic is an example of homeostasis. Ms. Norman explains that for this assignment, the videos are limited to just 60 seconds and may not be edited, so if mistakes are made, students will have to reshoot the video. She distributes a checklist and reviews it with the students to ensure they understand the assignment's requirements.

Ms. Norman also gives the class a QR code (a type of two-dimensional barcode) to scan with a QR code reader app that they previously installed on the personal mobile devices they've brought from home. The code takes students to a LiveBinder (a digital content collection at LiveBinders.com), featuring links they can use as they research their topic and write scripts for their videos. Students spend the remainder of the period reviewing the online resources provided by Ms. Norman and developing their scripts. Some students use a free journal app on their personal devices to take notes and write their scripts, while others opt for a traditional paper-and-pencil approach.

The next day, student groups practice and record their videos. They work in the classroom, the hallway, and the library media center. When students finish their videos, they upload them to a shared folder using Google Drive. During the last 20 minutes of class, Ms. Norman uses her laptop and projector to show the completed videos, and the class critiques them.

At a nearby elementary school, 5th grade teacher Paul Adams decides to mix things up a bit by incorporating use of mobile technology in a project-based lesson designed to teach students to write a character sketch. On the materials list, in addition to the six Android tablets and six Chromebooks permanently assigned to the classroom, are the students' own personal mobile devices.

On the first day of this multiday activity, Mr. Adams provides an overview of the project, explaining how authors use character sketches to improve their writing. He tells students that they will work in groups of four to create a fictional character and write a detailed description of this person. To help them get started, Mr. Adams shares links to 10 royalty-free photos of ordinary people from different walks of life. Each group selects one of the photos and uses an Android tablet to download their chosen image. They open the photo using a free app called Skitch, which allows them to add text and objects to the image. Today, they add the name and age of their character, identify at least six specific features in the photo that provide clues about the person, and use the app's arrow and text tools to call out each feature.

Over the next few days, Mr. Adams's students will use the information they brainstormed in Skitch to create a character map on the classroom's Android tablets using the free version of an app called Mindmeister. They will then use the classroom Chromebooks to write the character sketch, guided by a template stored on Google Drive. They will also use their personal devices as needed throughout this process, primarily for note taking and research.

1:1 Versus BYOD

Since the first desktop computers were installed in school computer labs, educators have struggled with how to optimize student access to technology. When most teachers had limited individual experience using personal computers, bringing students to the computer lab once or twice a week seemed to be a good solution. Over time, however, it has become clear that if technology is to have a real impact on learning, students need regular and immediate access to it.

The evolution of more portable laptop computers offered more flexibility in terms of where and how technology could be used. As the prices of these products dropped, schools and districts purchased more devices for use in classrooms. As early as the 1990s, educational institutions began experimenting with 1:1 initiatives, programs in which a school or district provides an Internet-connected device (typically a laptop in the early days, often a tablet now) to every enrolled student for the purpose of accessing digital instructional materials and the Internet. These 1:1 initiatives have lots of benefits. From an administrative perspective, the use of a common platform streamlines need for technological support and facilitates both equipment maintenance and software updates. In terms of equity, 1:1 means all students in the program have equal access to the technology. Then there is the point that monitoring students' use of the mobile devices is fairly simple, because all the devices are owned by the district or the school.

However, there are also drawbacks to the 1:1 approach, many of which are grounded in the need for sustainable funding to keep the initiative going. For example, initial investments in hardware, software or applications, subscriptions, and site infrastructure are high. There are also the ongoing costs of staffing, maintenance, repair, hardware replacement, subscriptions, and tools for monitoring, not to mention funding for regularly scheduled professional development opportunities for staff members.

Given the expense of successfully launching a 1:1 initiative and the growing number of students who own personal mobile devices (laptops, tablets, and smartphones), many schools are debuting Bring Your Own Device, or "BYOD," programs, where students and educators use personal technology devices at school. There are several reasons educators commonly cite for adopting BYOD. Because students express a preference for using their own devices whenever possible (Project Tomorrow, 2013a, p. 9), and a majority of parents report they are willing to purchase mobile devices for their children to use at school (Project Tomorrow, 2010, p. 12), BYOD programs can be less expensive for schools, both initially and on an ongoing basis. These cost savings enable schools and districts to reallocate limited financial resources toward expenditures for infrastructure or professional development. In addition, maintenance and equipment replacement costs become the parents' responsibility.

But BYOD programs present a number of challenges of their own. Students must remember to charge and bring their devices to school, and with those personal devices in hand, the potential for distraction is high. Teachers often find themselves reworking learning activities and classroom management strategies to ensure, as much as possible, that students who have continuous access to their social media feeds, games, and favorite means of communicating with friends stay on task. The technology can be a challenge, because even with defined minimum specifications for devices, teachers cannot assume students will be using a common platform. There is also the possibility that students won't have a device they can bring to school, raising concerns about equity. Further administrative concerns related to BYOD include the need to update policies and procedures to address such concerns as privacy, filtering content for appropriateness, and financial responsibility for required paid apps. And while the initial costs of BYOD programs may be less than those of 1:1 programs, they still present the ongoing costs of infrastructure maintenance and professional development.

Another issue brought to light by both 1:1 and BYOD programs is the fact that access to just one type of device—regardless of who provides it—is probably not enough. A report based on Speak Up Research Project results states, "Just as we do not assume that students will only access one book for all classes, the idea of using only one mobile tool to meet all assignment needs may be unrealistic" (Project Tomorrow, 2013a, p. 11). You may be aware of the 2013 Los Angeles Unified School District (LAUSD) rollout of a 1:1 initiative that would eventually have given an iPad to every K–12 student in the district. The project was fraught with problems on many levels (Lapowsky, 2015), but one concern that quickly rose to the surface was that iPads were not an effective one-device solution for LAUSD's high school students. My own experience working in schools with 1:1 or BYOD programs verifies this—if instructional needs are driving the selection of technology tools available at school, students will require access to more than one kind of device to complete various learning activities. Does this mean schools or parents must provide multiple devices for every student? No, but it does suggest that this may be a really good time to rethink what mobile technologies should be available to students and how to provide those technologies.

Some school districts are implementing hybrid mobile technology programs under which students may bring and use their own devices at school and use district-owned mobile devices such as tablets and laptops. These devices may be available for teacher checkout from a central location or permanently distributed to classrooms. The goal is not for the school to offer 1:1 access to one specific type of device but to make it possible for teachers and students to select the most appropriate tool for a given task from several readily accessible options. Some of these hybrid programs also include a component that allows students who do not own a personal device to check out one for use at school and often allows them to take the device home. Let's take a closer look at this type of initiative.

The Benefits of a Hybrid Mobile Technology Program

Is it really necessary for schools to provide mobile technology when so many students can bring their own devices? My experience working in schools both in the United States and abroad tells me that the answer to this question is yes—for both practical and pedagogical reasons. Implementing a hybrid mobile technology program requires schools or districts purchase fewer devices than does a 1:1 program, so hybrid programs are less expensive to initiate and maintain. Hybrid programs enable schools to offer access to multiple tools as opposed to just one kind of device. Depending on what technologies are purchased, it may be affordable to equip every classroom with several tablets and laptops or other devices. In those cases where this is not possible, schools often opt to invest in mobile carts shared by grade level or department. Use of shared, school-owned devices is also helpful when teachers want students to use a common platform for group activities or when teams need to have access to specific apps or programs that are installed on school-owned equipment. Hybrid programs can also free up funds for schools to purchase additional mobile devices for student check out to address equity concerns.

Hybrid programs offer further benefits in terms of pedagogical practice by expanding instructional options and potentially strengthening educational outcomes. For example, lessons that incorporate the use of a limited number of mobile devices facilitate small-group or pair work—an instructional mode that helps students develop valuable-communication and collaboration skills. This is important, because while it is common for educators to talk about engaging students in technology-infused activities that support higher-order thinking and 21st century skills, unless teachers *intentionally design* these kinds of learning modules, student endeavors in these areas will be limited at best.

The scenarios presented at the beginning of this book, depicting students working in hybrid mobile technology environments, illustrate how lessons incorporating the use of both personal mobile devices and shared ones support rigorous, reflective, and technology-supported learning. As you can see, the students have ready access to their own devices for more personal activities, like taking notes, but in each case they are also required to work collaboratively in groups with a shared device to create the final product.

A Few Words to the Wise

In this book, we will look at the practical logistics for launching and sustaining a successful hybrid mobile technology initiative. Before proceeding, I want to address concerns of readers who may be thinking, "But my site or district has already gone strictly 1:1 or BYOD. How can we shift gears now?" If the current initiative is working well for you and yielding the results you anticipated, no problem. Continue to monitor implementation, and plan to gradually incorporate aspects of a hybrid program that can help the initiative become even better.

But if things aren't turning out the way you'd hoped, now is the time to make adjustments that will address unanticipated consequences of your existing program. The research in support of a hybrid approach that is cited in this book is recent. We now know more about what effective mobile initiatives look like than we did even a year or two ago. We also know that new mobile technologies hit the market regularly, which means we won't be able to make a one-time decision about the technology in our schools and stick with it no matter what. Instead, we must build flexibility into every technology-supported initiative, recognizing that it's OK for the tools used to change as long as we keep our focus on sound education practices.

My experience in schools has convinced me that hybrid programs are a better, more flexible, and more practical approach to mobile device use in schools than the "either/or" of 1:1 or BYOD. I have also learned that there are definite keys to making a hybrid program work. Schools and districts must put in the time required for thoughtful planning, implementation, and evaluation. All parties must understand that simply deploying mobile devices will not automatically result in better learning outcomes for students.

If you set out to establish a hybrid program in your school or district—or set out to "hybridize" the 1:1 or BYOD program you currently have—almost invariably, two of the first planning challenges you will encounter will relate to the *state of the site or district infrastructure* and *the need for ongoing, high-quality professional development for staff.* Both of these topics will be discussed later in this book, but here are a couple of related caveats that will be deal breakers if they are not addressed right from the start.

The school or district's technology infrastructure must be able to handle any additional traffic that a hybrid program will generate. If the entire network cannot be upgraded at once, an alternative approach is to roll out the program in stages, adding grade levels or school sites as the capacity increases. A side benefit of this tactic is that the pioneering grade levels or sites can act as pilot programs, enabling educators to experiment and determine what does and doesn't work prior to fully executing the new program.

Appropriate, ongoing professional development is also a must. Teachers must become comfortable with the technologies provided by the school and should be prepared to make significant changes in classroom management strategies and activity design to accommodate the technology's use. If teachers are not given opportunities to learn and practice these changes prior to full implementation, a well-intentioned hybrid technology program may quickly devolve into students using mobile devices to complete activities that target the lowest levels of the revised Bloom's taxonomy (Overbaugh & Schultz, n.d.)—online quizzes testing recall, electronic fill-in-the-blank worksheets, books in PDF form, and so on. This is "technology-supported instruction" in name only, and it is not likely to impact student learning in any meaningful way. A further consequence is that students tend to stop bringing their personal devices to school because they view these activities as a waste of time.

Now, let's take a closer look at those areas your school or district can address to substantially increase the likelihood that a hybrid mobile technology program will work well for your students.

Planning for Success

A successful hybrid mobile technology program doesn't just happen; it is the result of careful planning and implementation, whether the effort is a new initiative implemented from the ground up or a shift from an exclusive BYOD or 1:1 approach. Unfortunately, there are several well-documented instances in which someone in a decision-making role was enticed by the glitz of one or more new technologies and then decided to push forward with a poorly conceived plan. While it is possible to launch a hybrid initiative with little or no buy-in or planning, these are the programs that are most often abandoned due to significant problems that arise almost immediately.

One of the most important steps in implementing a successful program is engaging members of the school or district community in the planning process. Begin by assembling a planning team whose members represent different points of view. Include all stakeholder groups: classified and certificated staff, administrators, students (this is especially important for middle and high schools), parents, and members of the community at large. You want team members who are technology users, but you also need to invite people with limited technology skills. This is a bit of a balancing act—you don't want a group that's too large to function, but you gain more credibility with a broadly representative team. Members of the team must be able to invest time to work together, resolve challenges, develop and implement a plan, and then participate in monitoring, evaluating, and revising that plan as needed. You may need to employ some creativity in meeting scheduling or other incentives to enlist the best people for this task.

It is important to remember that parental support is essential for the success of a hybrid program. In addition to including parent representatives on the planning team, you must keep the larger parent community informed about program development and implementation. These updates can be disseminated through the school or district's usual methods of communication, but it's also wise to offer open meetings where parents can learn more about the program and ask questions. Before such meetings, it may also be worthwhile to survey at least a sampling of parents to gain insight on their questions and concerns, and then tailor your agenda to provide the information or assurance they seek.

Here are some resources that provide additional guidance and pointers for establishing and working with a planning team:

- *Planning into Practice* (www.sedl.org/pubs/tec29/planning-into-practice.pdf)—Although it was published more than a decade ago, Chapter 2 of the SouthEast Initiatives Regional Technology in Education Consortium's *Planning into Practice* offers clear, easily implemented suggestions for forming and working with an advisory committee of this type (Sun, Heath, Byrom, Phlegar, & Dimock, 2000, pp. 19–50).
- *Why Create a Tech Advisory Committee?* (www.groundworkgroup.org/resources/tech-tips-archives/why-should-your-nonprofit-form-a-technology-committee)—GroundWork Group offers a downloadable presentation on forming an advisory group (De la Libertad, 2013).
- *What Makes an Effective Technology Committee in Education (v.2)* (www.speedofcreativity.org/2012/06/20/what-makes-an-effective-technology-committee-in-education-v-2)—Kent Brooks published an update to his blog post on what makes an effective technology committee.

During the initial meeting of the planning team, take time to start building solid working relationships. Set group norms and work collaboratively to establish a realistic time line. These steps are important—you want everyone to begin on the same page in terms of their commitment to the process. There are five key areas the group will need to consider during the process of planning and implementation:

1. Connection to curriculum
2. Infrastructure and support
3. Training and professional development
4. Budget
5. Policies and procedures

We'll talk about each in turn.

Connection to Curriculum

Infrastructure and professional development are important, but the planning process should start with the program's curricular connections. The primary reason for instituting any new program should be the potential for a positive impact on student learning, which means that all aspects of the hybrid mobile technology plan must have a clear connection to curricular goals and objectives. Too often, technology initiatives are launched with little or no thought about how well they will serve students. For example, one school district on the West Coast was poised to order nearly 1,000 tablet devices for middle school students to use as e-textbook readers, completely unaware that their adopted textbooks were not available in a format that would operate on that particular device. In another instance, officials in a Southwestern district were ready to launch a 1:1 program in grades K–5. Questions about what kinds of 1:1 activities would occur in kindergarten classrooms were met with blank stares—apparently, specifics about classroom instruction had not been discussed. Fortunately, in both cases, the districts delayed implementation until they could plan the programs more thoroughly.

You and your team can avoid close calls of this nature by first identifying students' learning needs and then determining if implementing a hybrid mobile technology program is an appropriate way to address those needs. Start by examining a recent overview of your school or district, such as a school improvement plan or school accountability report card. The overview should include background data that give a sense of the overall school or district climate, such as student demographics, enrollment, attendance summaries, discipline rates, academic performance summaries, school or district curricular goals, teachers' years of experience and education levels, and professional development plans.

It may also be beneficial to survey sample groups of staff, students, and parents about their perceptions of the school or district and their ideas about how students' educational experiences might be improved or enhanced, with or without use of technology (either the students' own or the school's).

This information is critical to program planning because it helps pinpoint where problem areas lie and how to effectively address such areas. Many factors influence student learning (National Center for Effective Schools Research and Development Foundation, 2003). For example, poor attendance rates or high percentages of suspensions and expulsions negatively affect student outcomes, because students are not learning when they won't or can't come to school. It's possible that redesigned learning environments that include the use of mobile technology devices would be an incentive for students to come to school and behave appropriately while there, but before adopting technology as the sole solution to such a problem, it is imperative that the team have a clear understanding of the current circumstances and probable causes.

Once you've identified existing issues related to student learning, brainstorm strategies to meet these needs. At this point, don't be put off by potential obstacles; explore ideas for an initiative you would implement in a perfect world. The odds are that instructional technology will come up in this discussion, because the majority of school community stakeholders believe that ready access to technology, particularly to mobile devices, would improve learning environments. Preliminary results from the 2015 Speak Up Research Project show that 75 percent of students in grades 6–12 and 51 percent of their parents want students to be permitted to use mobile devices in their classrooms. Educators agree: 77 percent of teachers surveyed think use of mobile technology increases student engagement, and 86 percent of school principals say it is important for students to have access to mobile technology in the classroom (Project Tomorrow, 2015). Note that mobile technologies are specifically cited in these survey results.

A next step is to discuss whether some type of hybrid mobile technology initiative would be suitable for the school or district. The discussion should generally address benefits of such a program, who would take part in the program, what kinds of activities students and teachers would engage in, when to launch the initiative, any specific concerns related to transition from an existing mobile device strategy, and how to measure the initiative's success. If a move to a hybrid mobile technology strategy involves expansion from an existing program, you will probably need to spend time educating the school community about why this shift is necessary, particularly if you've invested a lot of energy in "selling" the idea of a BYOD or 1:1 program. This can actually work in your favor if you use it as an opportunity to redirect the community's focus away from the specific tools students will use and toward students' learning outcomes. If the team decides that a hybrid mobile technology program is desirable, you should next review the two areas of immediate concern mentioned earlier: infrastructure and technical support, and professional development.

Infrastructure and Support

The current state of your school or district's technology infrastructure—both the network and the devices available for teacher and student use—will determine your next steps.

Until a short time ago, it was enough to ensure that the adults on campus had access to Internet-connected devices for work-related activities and that a computer lab or two, and perhaps library or media center workstations, were available for student use. These accommodations are no longer adequate. Today's educational environment requires that every student be able to connect to a reliable Wi-Fi network anywhere on campus. Unfortunately, many schools just aren't there yet.

Network and Connectivity Concerns

As recently as June 2014, nearly 60 percent of schools in the United States offered little or no Wi-Fi access to students (Wheeler, 2014). While it's true that this problem can be circumvented using 3G or 4G connectivity available to many mobile devices, districts often prohibit student use of this capability because it bypasses district network filters. Another concern is that these plans are fee-based, a potential equity hurdle for students whose families cannot afford monthly charges for personal devices. In addition, many schools or districts lack sufficient bandwidth.

Conduct an audit of the capabilities of the current school and district network, along with the potential for upgrades. The instructional technology department should be represented on the planning team; if it is not, bring in a staff member for all discussions related to infrastructure. Determine what is needed for every employee and student to be able to connect to the network anywhere on campus. If a full upgrade isn't possible immediately, what can be done right away, and how many students can be served by a limited upgrade? What is a realistic time line for a complete upgrade, and is funding available to pay for the work? Can the school or district roll out a hybrid program over time, as the infrastructure improves?

There are a number of readily available free planning guides you can use to conduct this audit (although these guides' titles reference 1:1 or BYOD programs, the data collection tools also work for hybrid initiatives):

- *1-to-1 Essentials Program* (www.commonsensemedia.org/educators/1to1)—This three-phase planning guide from Common Sense Education includes tools and suggestions.
- *K–12 Blueprint: BYOD* (www.k12blueprint.com/byod)—Sponsored by Intel Education, this guide offers planning tools, success stories, and additional resources.
- *One-to-One 2.0: Building on the "Bring Your Own Device" (BYOD) Revolution* (www.samsung.com/us/it_solutions/innovation-center/downloads/education/white_papers/One-to-One_2.0_-_Handbook.pdf)—Originally published by *Converge* magazine, this guide includes tips for determining your readiness for BYOD and helpful planning questions.

Additional planning guides may be accessed on the "Planning" tab of the "Bring Your Own Device (BYOD)" LiveBinder at www.livebinders.com/play/play?id=404796&backurl=/shelf/my.

The most common error at this stage is underestimating the demands that will be placed on the network when the hybrid program is launched. There are ways to work around connectivity issues for a short period of time, but most teachers and other staff working with students will need some type of training in these workarounds. Contact other schools or districts in your area that have successfully implemented schoolwide Wi-Fi for advice and support.

Device Selection, Distribution, and Use

Once network and connectivity concerns are addressed, the next step for the planning team is developing a vision statement for how mobile devices will be used to support student learning. This redirects your focus back to meeting students' learning needs and lays the groundwork effective program implementation.

Just as an audit of technology infrastructure is helpful for planning related to network concerns, an audit of the technology-supported instructional programs currently available to your students provides critical data for determining where things stand from a pedagogical perspective. The planning guides referenced above also include tools you can use to conduct this review.

Use the results of this audit to honestly assess your current instructional technology program—both its strengths and weaknesses. During the course of team discussions related to findings of this audit, you will want to address questions such as the following:

- What kinds of learning activities do students use mobile technology to engage in?
- When and where do students have access to school- or district-provided mobile devices?
- When and where are students permitted to use personal mobile devices?
- What kinds of technological devices do students use for learning activities?
- Is there a good match between learning activities and the tools available to do the work?

Your responses to these and other questions that arise in the course of discussion set the stage for the team to address the next important question: "How closely does current use of instructional technology mirror the vision statement for the use of mobile devices to support student learning?" Use your answer to launch an investigation into how to design a hybrid program to meet specific needs at your school or district. New questions to consider at this time include the following:

- What different kinds of mobile devices does the school or district need to provide?
- Who will select specific devices, and how will that decision be made?
- How many devices will be purchased?
- Where will the devices be located, and how will students access them?
- Who will be responsible for keeping track of these devices?

- How will these devices be maintained and repaired?
- What kinds of personal mobile devices are students going to be permitted to bring to and use on campus?

The last question in this list raises an issue that often unintentionally torpedoes BYOD initiatives. In an effort to be inclusive, educators are sometimes reluctant to establish minimum specifications for the devices that students bring to school. In reality, this is a necessary step. Students and their parents need a realistic understanding of required technological specifications so that they can plan to purchase an appropriate device. If students' devices need to have at least one camera and audio recording capability, let parents know. If a large amount of data storage space is required, head them off before they purchase something with just 8 gigabytes. Thinking through realistic minimal specifications helps determine how affordable recommended devices will be for families and also ensures that students will actually be able to use the devices they bring.

Your responses to the questions in this section aren't set in stone; think of them as guides that will help you draft a preliminary framework for your plan. Although further exploration is likely to lead you to reconsider some of the initial decisions you make, you must have an idea of how the program is taking shape in order to move the planning work forward.

Tech Support

In 1943, when Abraham Maslow proposed his theory of human motivation, he posited that if basic needs for food, clothing, and shelter are not met, a person will cease to function. In 2013, the author of a post on the OneXuan App blog adds two even more "fundamental needs" for modern society: charged batteries and Wi-Fi (Fonter, 2013). It's a tongue-in-cheek observation, certainly, but at the same time, insufficient tech support can be a root cause of failed technology initiatives. Schools and districts often deal with the BYOD part of a hybrid program by electing not to provide technical support for student-owned devices. But the network and devices owned by the school or district still need to be supported to protect your investment. How will this service be provided, and by whom?

Staffing levels for tech support in K–12 education are nearly always inadequate. The Consortium for School Networking (CoSN) conducts an annual survey of information technology (IT) leadership across the United States. The results of the 2014 survey show that although the number of devices to be managed increased in 77.4 percent of reporting districts compared to the previous year, there was a corresponding increase in staff in just 27 percent of reporting districts. In addition, more than 70 percent of survey participants reported being understaffed to the point that they cannot implement new technology or support integration of technology-supported instruction (CoSN, 2014b, pp. 23–25). In more specific terms, the ratio of technicians to students served across the United States is roughly one staff person for every 2,000 students (Schooldude, 2013).

In planning for a hybrid mobile technology program, your committee needs to consider the following tech support questions:

- How will you define the terms *maintenance* and *support*? What will providing these services entail?
- Do you have the necessary resources to maintain and support school or district technology? Where will these resources come from?
- How, specifically, will you provide support to end users—both educators and students?

The State Educational Technology Directors Association (SETDA) offers a free guide to implementing digital learning at http://digitallearning.setda.org/tech-support/#!/overview (SETDA, 2015). The guide includes a comprehensive section on technology support with 11 critical questions ranging from what tech support goals have been established to how tech support success will be measured, all of which are extremely useful when initiating conversations about the scope of tech support at your school or district. Other topics covered include strategies for procuring at least some technical support from vendors who sell equipment to the school or district, coordinating services provided by IT staff with professional development providers, and various ways to structure how technical support is provided to staff and students.

Training and Professional Development

Rapid changes in the technology landscape necessitate comprehensive professional development for teachers, both prior to launching an initiative and on an ongoing basis. Yet results of a 2012 survey of teachers across the United States showed that 82 percent of the respondents felt they had not received the professional development they needed to take full advantage of the instructional technology at their disposal (Hart Research Associates, 2012, Slide 22). A more recent survey conducted by Digedu Research Insights (2014) shows that things may be improving somewhat; nonetheless, 46 percent of respondents stated they do not receive adequate instruction on use of available devices, and more than 50 percent reported inadequate follow-up support for actual implementation of technology-supported instruction.

One major barrier to effective use of technology in instructional settings is the type of professional development offered. It's important to differentiate between *training* and *professional development*. For the purposes of this discussion, the goal of technology training is to increase an educator's rudimentary skills in use of one or more devices or digital tools. A well-*trained* educator understands operational basics. This is a good start, but more is required if teachers are going to design technology-supported learning activities that have a positive impact on student learning. This is where professional development comes in—it goes beyond fundamental skills, providing educators opportunities to learn how they can use the technologies to support instruction beyond simple task automation—and truly leverage the technologies' capabilities in ways that enhance and expand learning opportunities for students.

Take a look at any technology-related professional development your teachers have attended recently. How much time was spent focusing on basics, such as how to operate a device or how to use a particular digital tool? How much time was spent learning to use technology to manage administrative tasks or to automate classroom activities like drill and practice, pop quizzes, or conducting research? Finally, how much time was devoted to ways technologies can be used to encourage students to use higher-order thinking skills to analyze, evaluate, or create something? In most cases, such courses tend to be heavy on the training side and limited as to actual professional development. The solution is to develop a plan that provides opportunities for *both* training and professional development.

Technology Training

In a hybrid mobile technology setting, educators will need to understand how to operate the devices being provided by the school. Typically, schools using hybrid programs acquire two or three kinds of devices for student use.

Depending on how the equipment is distributed, teachers may have ready access to one or more of these devices. For example, an elementary school might have 5 tablets in every classroom in grades K–2, 10 Chromebooks in every classroom in grades 3–5, and 30 tablets available for check out in groups of 5 to grades 3–5. Initial training for primary teachers would focus on how to operate the tablets. Upper elementary teachers would need to learn how to use both the Chromebooks and the tablets. At the high school level, a hybrid program might provide laptops or convertible notebooks (devices that can be used as a laptop or tablet) for student use in classrooms and the library media center, using cloud storage for student work. All teachers would need to know how to operate the laptops or convertible notebooks and be able to help students access and save their work in the cloud.

What about the devices students bring from home? In many cases, the hybrid plan will state that teachers are not responsible for troubleshooting or otherwise helping students use personal devices. In reality, teachers almost always end up offering minimal assistance to students who are struggling. This is one reason it benefits educators to have a basic familiarity with the features of the devices that students bring to school, but there is a more compelling one. Knowledge of the basic capabilities of various devices helps teachers develop lesson plans that support effective technology use.

It's a good idea for teachers to begin each semester by surveying students to determine which devices and which operating systems they will be bringing to school (smartphone, tablet; iOS, Android, Windows). This doesn't mean that every device in the classroom needs to be in use at all times; in fact, there are good reasons to encourage students to share devices and work collaboratively. That said, having a general sense of what devices are being brought by students and what features might work well in specific activities will make planning easier for teachers, and it demonstrates to students and parents that the BYOD part of the hybrid initiative is being taken seriously.

The previous paragraph underscores why it is important to establish minimum specifications for the devices students are permitted to bring to school for use in class. Of course there will be differences from one device to the next, but basic technology training that addresses common aspects of most mobile technologies will help prepare teachers to troubleshoot problems with student-owned devices. Consider developing a simple troubleshooting protocol that teachers can use themselves and share with students. Here's an example:

BEFORE YOU ASK FOR HELP:

1. Force close the unresponsive app.

2. Check the Wi-Fi connection.

3. Reboot the device.

These three steps will work on virtually any device and will take care of many operational problems. If they don't resolve the issue, the device can be put aside and dealt with later, following the procedures established for that type of device.

Let's look at a few other topics to cover in basic training for teachers.

Wi-Fi requirements. Of course Wi-Fi is required to download apps wirelessly, but once installed, not all apps require a wireless connection to work. Most teachers know this is true for software used on laptops but don't realize it can also apply to tablets, smartphones, and even some uses of Chromebooks (Google, 2015). The ability to use mobile devices offline is particularly helpful when the school's network may not be as robust as desired. Of course, students will eventually need to connect to the Internet to upload their work to the cloud, but that can be done at home (for student-owned devices) or outside the school day (for school-owned devices). You can determine which apps work without an Internet connection by turning off Wi-Fi on your device and attempting to run the software. Some apps that lend themselves to offline use include camera software (pre-installed on tablets and smartphones); MindMeister (Web, iOS, Android); Skitch (Web, iOS, Android, Windows 8.1); Book Creator (iOS, Android); and Explain Everything (iOS, Android), as well as Google Docs, Sheets, and Slides (Web, iOS, Android).

Device-agnostic apps. The list of apps shared above brings up another topic for basic training. Notice that every app on the list runs on at least two platforms. At one time, many apps were limited to just one device; however, this is no longer the case. Some education technology experts call cross-platform apps "device agnostic." Educators who use apps that run on more than one platform will find they have an easier time planning learning activities because they won't have to take compatibility with various devices into consideration. By the same token, in those cases when it's important to use an app that is device-specific, teachers can be aware of this ahead of time and make sure they will have the appropriate device available for student use by either checking out compatible school-owned devices or asking students who own compatible devices to download the app (if free) and bring it to school on the appointed day (for paid apps, follow school or district policies for downloading paid apps to student devices). Educators can determine which platforms are supported by an app by visiting the app publisher's website or by conducting an Internet search using the name of the app along with a specific platform (e.g., Android or Windows). You may want to create a Google Doc to build a list of device-agnostic apps (including the supported platforms) for sharing with colleagues on site or in the district.

Equivalent apps. Although many apps are now supported by multiple platforms, there are times when this is not the case. For example, iTalk is an easy-to-use app for audio recording on iOS devices, but the publisher does not offer versions for other platforms. It is often possible to find an equivalent app for other platforms by conducting an Internet search using keywords that describe the desired function. In the example provided here, a search using the keywords "audio recording apps" will provide alternative apps for Android and Windows operating systems.

These suggestions are fundamental skills educators need to be trained in so that they can identify devices and digital tools for use in their classrooms. However, making final selections for instructional use requires more advanced skills that they can acquire through professional development activities that focus on instructional design rather than basic operations.

Professional Development in Technology-Infused Instruction

A number of years ago, Alan November (1997) argued that instructional technology has no significant impact on student learning when it's used just to automate what students would do anyway, with or without the technology. He proposed that when instructional technology use is focused on information and communication—he called this informating—educators can get different results for students and their families. Starting with the Apple Classrooms of Tomorrow (ACOT) report published in 1995, two decades of research confirm this idea, yet teachers are still mostly using technology to manage administrative tasks, not to support instruction (Apple Computer, Inc., 1995; Project Tomorrow, 2013a).

To ensure that your hybrid program effectively supports teaching and learning, it's important to rectify this situation, but where should you begin? There are three readily available resources educators can use as a foundation for designing professional development activities that will actually help participants make better use of instructional technology.

The SAMR model. The Substitution Augmentation Modification Redefinition (SAMR) model is a four-step, two-phase map designed to assist educators who want to incorporate effective use of technology into instruction (Department of Education and Training, 2013). Developed in 2010 by Dr. Ruben Puentedura, the SAMR model is reminiscent of ACOT's Stages of Technology Use (Apple Computer, Inc., 1995) and the Levels of Teaching Innovation Framework (LoTi) developed by Chris Moersch in 1994 (Bishop, n.d.). These similarities lend credibility to the SAMR model, which reflects ideas about technology integration that are supported by previous research and have withstood the test of time.

The SAMR model depicts how instructional technology use develops in stages from simple to quite complex. This model's two-part approach also affirms that earlier stages of use have less impact on students' academic performance than later stages, during which students are challenged to engage in more complex tasks. This does not mean that it is recommended or even desirable for every technology-supported activity to advance through all four steps. What is important is that educators are able to identify the level of technology use represented in an activity and articulate how it reflects what the teacher meant to accomplish in terms of academic goals. In addition to considering the level of sophistication for any given task, teachers can use SAMR stages to assess their own personal levels of use of various technologies. Incorporating this model into professional development activities helps teachers become better at designing appropriately complex technology-supported activities.

The revised Bloom's taxonomy. The original Bloom's taxonomy was developed in 1956 to help classroom teachers incorporate higher-order thinking skills into learning activities on a regular basis. Beginning in the mid-1990s, Lorin Anderson and David Krathwohl convened a group to update the taxonomy, and the revised Bloom's taxonomy was published in 2000. Major revisions included changing level names from nouns to verbs to reflect the idea that thinking is active, and flipping the top two levels to place creating at the pinnacle of the taxonomy pyramid (Clark, 2015).

It is important for educators to consider the revised Bloom's taxonomy when planning technology-supported instruction because all too often classroom use of any form of technology is confined to activities that target the first two levels of the taxonomy—*remembering* and *understanding*. Examples include online quizzes and drill-and-practice games. There are certainly times when students need to work at these levels; however, educators need to engage in professional development activities designed to help them design technology activities that target higher levels of the taxonomy—*applying, analyzing, evaluating,* and *creating*.

A related resource to share with teachers is Bloom's digital taxonomy, a diagram created by Andrew Churches (2015) that illustrates the alignment between each level of the revised Bloom's taxonomy and digital tools that support thinking skills at that level. The illustration is very useful for teachers who mistakenly think, for example, that having students conduct online research is all that's required to fulfill their responsibility to incorporate technology use into learning activities.

The TPACK framework. Once educators are familiar with the SAMR model and the revised Bloom's taxonomy, it's very helpful to share the Technological Pedagogical Content Knowledge (TPACK) framework, developed by Dr. Punya Mishra and Dr. Matthew J. Koehler of Michigan State University (Koehler, 2012). The TPACK framework helps educators understand the types of technology activities that can be used at various levels of the SAMR model and the revised Bloom's taxonomy.

Most teachers realize that generating well-designed traditional instructional activities requires a clear grasp of the content to be taught and familiarity with multiple instructional strategies (pedagogy) so that they can identify and use the most appropriate strategy. Think of "pedagogy knowledge" and "content knowledge" as circles placed side-by-side so they intersect—a Venn diagram. The goal in designing traditional lessons is to develop activities that fall within the intersection. But technology-supported instruction requires more.

The third element required for designing technology-supported lessons is knowledge about how to use the technology tools available. When this third circle ("technological knowledge") is added, the Venn diagram becomes far more complex. Take a look at the TPACK image in Figure 1. As you see, instead of just one intersection, there are three places where two of the three circles cross one another, and all three circles intersect at the center of the diagram. The goal in good design of technology-supported lessons is to design activities that incorporate all three knowledge areas. This is also where the challenge lies for most teachers. It's fairly easy to include two of these areas, but incorporating all three can be more difficult.

FIGURE 1: **The TPACK Framework**

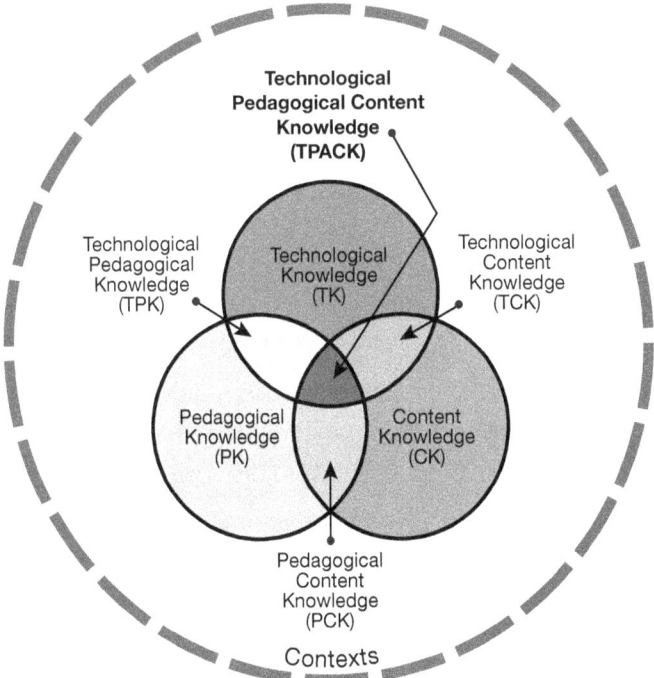

Reproduced with permission of the publisher, © 2012 by tpack.org

Fortunately, there are many online resources related to the TPACK framework to help teachers design effective technology-supported learning activities, including the following:

- TPACK.org (www.tpack.org)—Maintained by Matthew J. Koehler, this site offers and overview of the TPACK framework, links to articles about TPACK, and a forum where educators can post questions about TPACK.
- Learning Activity Types Web Site (http://activitytypes.wm.edu)—Hosted by the College of William and Mary School of Education, this website features ideas for "operationalizing" TPACK. Content area links provided in the left sidebar lead to tables listing suggested instructional strategies and technologies ranging from entry-level activities through undertakings that require higher-order thinking skills. Although somewhat dated, the site is still an excellent resource for K–12 teachers.
- The TPACK Game (www.matt-koehler.com/the-tpack-game)—One of my favorite TPACK sites, this interactive game provides practice in thinking about how the three knowledge areas interact in designing effective technology-supported instruction. The game randomly selects two of the three knowledge areas, and players work together to define what the missing knowledge area could be. For example, if the content area is expository writing and the pedagogy is discussion/debate, what technology could be used to support this activity?

These sites and other resources are available to readers in a Bag the Web collection called Resources: TPACK Framework (www.bagtheweb.com/b/MQLRej).

Professional Development in Classroom Management

In conjunction with learning about new approaches to instructional design using technology, educators need professional development in adopting and adapting new classroom management practices. Classrooms operate differently when students are using personal and school-owned mobile devices. I recently observed a middle school classroom where a young teacher permitted students to use personal mobile devices for an activity but didn't change any of the classroom management strategies he normally used. Students were sitting in pairs at narrow tables arranged in rows facing the front of the classroom. It was difficult to move around the room freely to check in with individual students, so the teacher spent most of his time lecturing from the front of the classroom. Occasionally, he asked students to use their devices to look up vocabulary definitions. When the teacher directed students to place their mobile devices on the table, he neglected to ask them to put all of their devices there. As a result, several students had second devices in their laps and spent much of the lesson texting—something that went unnoticed by the teacher at the front of the room.

Examples like this one make it readily apparent that incorporating mobile technology devices into the classroom requires adapting both instructional practice and management techniques. Discussions of how to change classroom management practices are more fruitful in the context of lesson plans teachers have developed using the models we have discussed. Without the framework of an actual lesson, these conversations tend to be theoretical and generalized. Here are some suggestions for online resources for strategies educators can use with their students as they implement technology-supported instruction:

- *BYOD Teacher Management Tips* (http://bit.ly/1wGDfOr)—Posted by Burlington-Edison School District in Washington state, this is a compilation of three tips lists.
- *Classroom Management* (https://sites.google.com/a/whps.org/byod/ideas-for-the-classroom)—Posted by West Hartford Public Schools in Connecticut, this page offers several broad suggestions and then six specific tips.
- *10 Classroom Management Tips for BYOD* (http://bit.ly/1RYF7NJ) and *10 More Classroom Management Tips for BYOD* (http://bit.ly/1HElZVO)—The 20 tips here come from a Canadian blog based in Burlington, Ontario. The first list of tips is similar to other resources mentioned here. The second list of tips includes ideas for ways to use devices as well as manage them.

Combining Training and Professional Development

Once a robust infrastructure is in place, it is the combination of teacher training and professional development that forms the heart of a successful hybrid mobile technology initiative. Your school or district can create an environment that supports administrators, teachers, and support staff as they strive to achieve the learning experiences they require for successful implementation. Just as a curricular needs assessment and technology audit are invaluable for determining next steps in these areas, it is important to take time to review existing training and professional development offerings to see if they will be sufficient. Here are a few questions to ask:

- Do all staff members have the basic skills required to use the devices provided by the school or district? If not, what basic training topics need to be covered?
- Are staff members familiar with models for effective technology-supported instructional design? If not, what are the gaps and how will these needs be addressed?
- How will basic training and professional development be differentiated to meet individual staff needs?
- What kinds of follow-up support are currently offered? Is it feasible to provide in-class coaching or ongoing professional learning communities to support implementation of a hybrid program?
- What digital instructional materials are available to support the curriculum, and which of these are currently in use?
- How well do curricular activities encourage the use of higher-order thinking skills?

Use the responses to these and other questions developed by your team to evaluate your professional development needs. Identify existing gaps in the current program and determine how to fill them, both before and during program implementation. Draft a plan and calendar for the school year that includes both basic training and professional development sessions. Determine who will provide training and professional development for staff.

Depending on the background and experience of site or district IT staff or other knowledgeable staff, it may not be necessary to bring in an outside provider for basic technology training. The advantages of in-house training are that current employees are more familiar with existing infrastructure and devices, and they may also be available to provide ongoing support when questions and challenges arise. Access to technological support when it's needed is nearly as important as access to high-quality training experiences.

Professional development may also come from within the school or district itself; however, it may be helpful to contract with outside consultants for at least some of these offerings, depending on the topic. The benefit of in-house providers is that they will be familiar with school or district policies and procedures, which will need to be incorporated into professional development experiences. In addition, in-house staff can deliver ongoing support in and outside of the classroom more readily than a consultant. On the other hand, consultants bring experiences and expertise that internal employees may lack. An ideal solution may be to find one or more consultants who will work closely with internal professional development providers and other staff members to ensure continuity and a consistent message.

Budget

Several years ago, technology budget conversations nearly always included a discussion of the Total Cost of Ownership (TCO) of devices, infrastructure, and so on—a comprehensive calculation of how much a program would cost both up front and over time, including such expenses as hardware, infrastructure, technical support, utilities, and ongoing professional development. Although the term *TCO* is no longer used as frequently, the fact remains that technology usually costs more than we realize. For example, 1:1 initiatives are very expensive, initially and over time, thanks to immediate and ongoing expenses related to keeping the program afloat. BYOD programs are usually less expensive, because families assume responsibility for purchasing and maintaining the technology that students use at school, but costs are still incurred. Hybrid mobile technology initiatives fall somewhere in between, depending on the kinds of technologies provided by the school or district and the number of devices purchased.

The important thing to remember is that there are certain initial and ongoing costs that schools and districts must absorb, regardless of which strategy is adopted. These costs include but are not limited to infrastructure, repair and maintenance, IT staffing, training and professional development, and digital instructional materials. Technology budgets also include ongoing costs, and thus need to be developed for multiple years to ensure program sustainability.

When it comes to budgeting for technology expenditures, stable funding sources that are available over time (e.g., general funds) are preferable to categorical or grant dollars tied to special programs or projects, which often have a defined life span. This is particularly important when ongoing expenses, such as salaries and app or software subscriptions, are involved. The *K–12 Blueprint: Funding* toolkit (www.k12blueprint.com/funding) offers a comprehensive overview of budget planning for technology initiatives including funding models, ideas for fundraising, and more. The U.S. Department of Education's Office of Educational Technology has also developed a quick reference guide educators can use for infrastructure planning that includes questions related to budget and potential funding sources (http://tech.ed.gov/files/2015/02/Infrastructure-Quick-Reference.pdf).

Let's look at a few strategies that might be used to augment insufficient funding.

E-rate

The Consortium for School Networking (CoSN) identifies lack of funding as the most significant barrier to sufficient connectivity and robust technological infrastructure in schools (CoSN, 2014a). This is problematic because the scale of a new hybrid program will be defined by the current capabilities of the existing network, and funding for upgrades today and in the future will probably be an issue.

One funding source to look into is the E-rate program sponsored by the Federal Communication Commission (FCC). As of July 2014, E-rate funds are available to schools and libraries to provide students with reliable Wi-Fi connectivity (FCC, 2014). These dollars are currently planned to be available until 2019.

Fundraising

Fundraising—acquiring voluntary contributions of cash to pay for a specific project or program—is a way to pay for one-time expenditures, but it is not a great strategy for supporting ongoing costs. Identifying a specific goal with a deadline that can be tracked over time is attractive to potential contributors. Recent years have seen the appearance of websites that support online fundraising for individual teachers and schools, including the following:

- DonorsChoose.org (www.donorschoose.org)—Public school teachers in the United States post requests on this site, indicating how much money is needed to fully fund their project. Requests may be posted for up to four months. Donors review projects and decide how much they will contribute to one or more projects. If a funding goal is not reached in time, donors may redirect their pledge to another project or send the teacher a DonorsChoose.org gift card for the pledged amount.
- AdoptAClassroom.org (www.adoptaclassroom.org)—Teachers at public and private schools across the United States may register on this website to generate

wish lists for a wide variety of classroom needs, including technology. Donors make contributions that receiving teachers can use to purchase items through the site.

Of course, there are many ways to raise funds for school needs. One principal at a private school in Southern California came up with a novel approach. Concerned about equal access to mobile devices for students who couldn't afford to purchase their own, she set out to find a way to make the hybrid program feasible for the families in her school. The budget simply did not allow for purchasing devices for home use outright, so she offered to work with the school's parent organization to host fundraising activities, such as walk-a-thons and car washes, throughout the summer. All proceeds from these fundraisers were pooled into one account, and every student who participated fully in the event was credited with an equal share of the money earned (specific guidelines about what constituted full participation were in place). By the end of the summer, every child who stuck with the program had earned a device that was purchased with these funds and given to the student. Parents of students who did not earn the full amount needed could pay the difference, or the credit was returned to the general account to be redistributed. Although these events took place in a private school, public schools should be able to come up with a version of this idea that would work for them.

In-Kind Donations

In-kind donations are another way to meet budget shortfalls, providing a way to acquire tangible items such as hardware or instructional materials. Businesses and other organizations often donate used computers and other technology equipment to schools or districts. While this practice can be a boon to cash-strapped schools and districts, used technology equipment often ends up becoming a financial burden for the recipient. For example, it may be necessary to pay to have the equipment refurbished before it can be used, or worse yet, the beneficiary may end up having to pay to properly dispose of unusable donated devices. One way to deal with well-meant gifts of used equipment is to establish guidelines for accepting them. For example, the school or district can set minimum specifications for computers, tablets, smartphones, printers, and other devices. Equipment that does not meet the threshold described can be politely refused.

New equipment and instructional materials may also be donated, which is usually preferable to donations of used items. As with cash solicitations, there are websites designed to facilitate donations of new items to schools. Here are a couple of examples:

- Digital Wish (www.digitalwish.com)—This website is specifically designed for technology donations. Teachers in public and private schools post wish lists for equipment and/or instructional materials. Donors browse a database of teacher requests from around the United States, select a teacher to support, and purchase an item from the teacher's wish list. The item(s) purchased are shipped directly to the recipient's school.

- ClassWish (classwish.org)—Teachers set up wish lists for items they need for their classrooms, including technology. Donors select a teacher to support and make a contribution. Teachers use their wish list account to purchase items through ClassWish.

Sustainability

Launching a hybrid mobile technology program is one thing; maintaining it is something else. Schools and districts often struggle with keeping an initiative alive once the original funding source has dried up. I know of two districts that opted to fund 3G/4G connectivity for district-owned student devices during the first year of an initiative so they could buy time for network upgrades. In one case, the district obtained a grant to cover related costs for the upgrades, and the stopgap strategy worked. In Year Two, the district was able to rely on Wi-Fi connectivity and drop the 3G/4G plan. In the second case, funding for connectivity plans dried up after Year One, and the network still was not upgraded by Year Two. This resulted in limited use of the devices once 3G/4G connectivity was no longer available.

In these examples, paying for 3G/4G connectivity for one year resulted in budgets that were balanced for that fiscal year, but neither approach actually reflected a good overall spending plan. The first district did manage to find additional funding to pay for upgrades, but the second district did not. All too often, schools and districts develop budgets that are balanced in the short term but rely on grants or other funding sources for long-term projects that may or may not materialize. When the funds don't come through, projects end up being postponed—sometimes indefinitely.

Here are a few suggestions for creating more sustainable budgets:

- Spend at least as much time analyzing a proposed budget as creating it.
- Make certain that the proposed budget accurately reflects the priorities of the school or district plan.
- Refer to the approved budget as a guide for all expenditures.
- Use current funding levels and past funding trends to forecast future budgets over several years, understanding that funding amounts change annually.

Policies and Procedures

Writing policies and procedures for your hybrid mobile technology initiative offers another opportunity for the planning team to think through specifics related to how the initiative will play out in the classroom and establishes the tone for the program. Policies are the documents that describe broad principles, rules, and guidelines to meet stated district or school goals. Procedures are the documents that provide specific steps for policy implementation.

It's important that expectations and implementation steps are clearly stated, without becoming so detailed and restrictive that educators, students, and parents are overwhelmed. Look for times when existing policies and procedures can be used to support the hybrid initiative to avoid redundancy. For example, consider referencing the current Code of Conduct for disciplinary measures rather than developing a whole new set of rules for student behavior under the hybrid program.

Some schools or districts decide not to develop policies and procedures prior to implementing new programs, instead adopting a "wait and see" attitude. I am not aware of any occasion when this approach worked out well. It is better to develop a policy for a hybrid program that lays out a framework, even if it may need future revisions, than to begin with nothing in place. Let's take a quick look at some of topics and questions to consider as you create a policy for your program.

Program overview. What is the purpose and philosophy behind the hybrid initiative, and how does it differ from 1:1 and traditional BYOD programs? Will the hybrid program provide an opportunity for students to learn associated skills, such as personal responsibility, communication, and collaboration? When will implementation commence? Which students will participate?

Acceptable devices. What kinds of devices will students be permitted to bring to school? Will there be a list of minimal specifications or specific devices required? What kinds of devices will the school or district provide to ensure access to a common platform when needed? Will there be a process for approving or registering personal devices used at school? How will equity issues be dealt with for families who cannot afford to purchase a personal device?

Software/apps. How will educators determine what software or apps they will use to support instruction? Will district- or site-owned software and apps be installed on personal devices? If a teacher wants students to use a paid app or piece of software on personal devices, how will that be handled? Will it be permissible to ask students to download free apps to their own devices? Who is responsible for downloading and installing software and apps on personal devices?

Connecting to the network. How will students access the network? What filters are in place to block inappropriate sites and comply with federal mandates (e.g., Children's Internet Protection Act or Children's Online Privacy Protection Act)? How will students' online activity be monitored? May students use personal 3G/4G data plans? If so, who is responsible for associated fees, and how will students' 3G/4G use be monitored?

Security and care of devices. Who assumes responsibility for repairs, theft, or damage to personal devices? Who is responsible for keeping batteries charged? May teachers provide charging stations in their classrooms? Will district employees troubleshoot student-owned devices that are not working? If so, to what extent?

Acceptable use policy. This final area for consideration gets a bit tricky. Some schools and districts maintain separate acceptable use policies (AUPs) for adults and students, while some opt to integrate the AUP into the overall hybrid program policy. Regardless of the approach taken, here are questions to explore while developing a policy on the use of devices:

- How will student-owned devices be used to support classroom activities?
- How will school or district-owned devices be used to support classroom activities?
- Are there restrictions for the kinds of digital tools teachers can incorporate into instruction, and if so, what are they?
- Are students permitted to use personal devices anywhere on campus? If not, where may the devices be used?
- Are students permitted to use their own devices while on campus to engage in noninstructional activities, including texting, using social media, or personal web browsing? If so, when and where?
- How will behaviors related to digital citizenship (e.g., cyberbullying, copyright, plagiarism) be addressed? Are existing policies and procedures related to student behavior applicable for the hybrid program as well?

A well-written, concise policy needs to be supported by procedures that spell out how to carry out rules and guidelines found in that policy. Without clear, manageable, and realistic procedures to follow, teachers and students can flounder and inhibit the success of the hybrid mobile technology program. Once the program policy is drafted, the planning team should review each section to determine where additional support will be required in the form of specific procedures.

* * *

I hope your school or district will consider adopting a hybrid approach to incorporating mobile technology devices into classroom instruction. Although a program of this type requires a great deal of time to properly plan and implement, it can be a game changer for educators and students alike.

To give your feedback on this publication and be entered into a drawing for a free ASCD Arias e-book, please visit
www.ascd.org/ariasfeedback

ASCD | arias

ENCORE

PLANNING AND IMPLEMENTATION CHECKLIST

Use this checklist as a guide to launch your hybrid mobile technology program.

Plan for Success
- [ ] Establish a planning team that represents all stakeholder groups.
- [ ] Set group norms and expectations.
- [ ] Develop a realistic planning time line.
- [ ] Design a meeting schedule that encourages attendance by all members.
- [ ] Communicate with parents to garner their support for the program.

Connection to Curriculum
- [ ] Review recent data to get a sense of overall school or district climate.
- [ ] Survey stakeholders to solicit their ideas for improving students' education experiences, with or without use of technology.
- [ ] Identify existing issues related to student learning.
- [ ] Discuss multiple strategies that might be adopted to address issues related to student learning.
- [ ] Determine if it is appropriate to pursue a hybrid mobile technology program.

Infrastructure and Support
- ☐ Conduct an audit of the current school and district network infrastructure, along with the potential for upgrades.
- ☐ Develop an upgrade plan that will allow every employee and student to connect to the network anywhere on campus.
- ☐ Establish a realistic time line for the upgrade plan.
- ☐ Develop a vision statement regarding how mobile devices will be used to support student learning.
- ☐ Conduct an audit of the technology-supported instructional programs currently available to your students.
- ☐ Identify gaps between current programs and your vision for using mobile technology devices to support student learning.
- ☐ Create an initial draft of a framework for your hybrid program.
- ☐ Define the terms *maintenance* and *support*, and describe what providing these services entails.
- ☐ Determine if you have the necessary resources to maintain and support school or district technology.
- ☐ Make specific decisions about how support will be provided to end users (educators and students).

Training and Professional Development
- ☐ Review existing training and professional development offerings for staff to determine if current offerings

will be sufficient for the hybrid mobile technology program.
- [ ] Identify gaps between existing training and professional development offerings for staff and your vision for using mobile technology devices to support student learning.
- [ ] Draft a plan and calendar for the school year, including topics for basic training and professional development, and ascertain who will provide these services.
- [ ] Determine if you have necessary resources to offer appropriate training and professional development for staff.
- [ ] Make specific decisions about how ongoing support will be provided to educators.

Budget
- [ ] Use current funding levels and past funding trends to forecast future budgets over several years, understanding that funding amounts change annually.
- [ ] Develop a multiyear budget that includes a plan for sustainability.
- [ ] Spend at least as much time analyzing the proposed budget as creating it.
- [ ] Make certain that the proposed budget accurately reflects the priorities of the school or district plan.
- [ ] Regularly refer to the approved budget as a guide for all expenditures.

Policies and Procedures
- [ ] Investigate mandated procedures for developing and approving policies and procedures for your school or district.

Follow steps below within local guidelines:
- [ ] Review existing policies and procedures to determine if they can be used to support the hybrid initiative.
- [ ] Draft a policy for the hybrid mobile technology program that clearly identifies guidelines, expectations, and implementation steps.
- [ ] Review guiding questions for acceptable use policies and draft an AUP for your program.
- [ ] Follow established guidelines for reviewing, editing, and obtaining approval for policies and procedures.

References

Apple Computer, Inc. (1995). Changing the conversation about teaching, learning, & technology. Retrieved from http://imet.csus.edu/imet1/baeza/pdf%20files/upload/10yr.pdf

Bishop, B. L. (n.d.). Moersch's LoTi hierarchy for technology integration. Retrieved from https://sites.google.com/a/garesa.org/usingtechdeeply/home/moerschloti

Churches, A. (2015). Bloom's digital taxonomy. Retrieved from http://edorigami.wikispaces.com/Bloom%27s+Digital+Taxonomy

Clark, D. (2015). Bloom's taxonomy of learning domains. Retrieved from http://www.nwlink.com/~donclark/hrd/bloom.html

Consortium for School Networking (CoSN). (2014a). K–12 report: Affordability, adequate funding biggest technology barriers. Retrieved from http://cosn.org/about/news/k-12-report-affordability-adequate-funding-biggest-technology-barriers

Consortium for School Networking (CoSN). (2014b). 2014 K–12 IT leadership survey. Retrieved from http://www.cosn.org/sites/default/files/pdf/IT%20Leadership%20Survey%20Master%20Report_Final_1_0.pdf.

De la Libertad, B. (2013). Why create a tech advisory committee? Retrieved from http://www.groundworkgroup.org/wp-content/uploads/2014/03/Creating-Your-ITCommitteeFeb2014.pdf

Department of Education and Training. (2013). The SAMR model: Engage in deep learning and authentic contexts. Retrieved from https://classroomconnections.eq.edu.au/topics/pages/2013/issue-7/samr-learning-technologies.aspx

Digedu Research Insights. (2014). Technology use in the classroom: Benefits and barriers. Retrieved from https://www.youtube.com/watch?v=ru82F07TKA0

Federal Communications Commission. (2014). FCC modernizes E-rate program to expand robust Wi-Fi networks in the nation's schools and libraries [press release]. Retrieved from https://www.fcc.gov/document/fcc-modernizes-e-rate-expand-robust-wi-fi-schools-libraries

Fonter. (2013). New Maslow's hierarchy of needs [blog post]. Retrieved from http://onexuan.com/blog/2013/09/new-maslows-hierarchy-of-needs/

Google. (2015). Use your Chromebook offline. Retrieved from https://support.google.com/chromebook/answer/3214688?hl=en

Hart Research Associates. (2012). Parents' and teachers' attitudes and opinions on technology in education: Key findings from quantitative research conducted August 2012 for the LEAD Commission. Retrieved from https://app.box.com/s/nfpkody26rx9prhyqvcd

Koehler, M. (2012). TPACK explained. Retrieved from http://matt-koehler.com/tpack/tpack-explained/

Lapowsky, I. (2015). What schools must learn from LA's iPad debacle. Retrieved from http://www.wired.com/2015/05/los-angeles-edtech/

National Center for Effective Schools Research and Development Foundation. (2003). Effective schools research base / Donnelley and Lee Library Archives and Special Collections at Lake Forest. Retrieved from https://www.lakeforest.edu/library/archives/effective-schools/EffectiveSchoolsResearchbase.php

November, A. (1997). Creating a new culture of teaching and learning. Retrieved from http://novemberlearning.com/educational-resources-for-educators/teaching-and-learning-articles/new-culture/

Office of Educational Technology. (2015). Future ready schools: Quick reference guide of key questions for planning technology infrastructure. Retrieved from http://tech.ed.gov/files/2015/02/Infrastructure-Quick-Reference.pdf

Overbaugh, R. C., & Schultz, L. (n.d.). Bloom's taxonomy. Retrieved from http://ww2.odu.edu/educ/roverbau/Bloom/blooms_taxonomy.htm

Project Tomorrow. (2010). Creating our future: Students speak up about their vision for 21st century learning: Speak Up 2009 National Findings. Retrieved from http://www.tomorrow.org/speakup/pdfs/SU09NationalFindingsStudents&Parents.pdf

Project Tomorrow. (2013a). From chalkboards to tablets: The emergence of the K–12 digital learner: Speak Up 2012 National Findings. Retrieved from http://www.tomorrow.org/Speakup/Pdfs/SU12-Students.pdf

Project Tomorrow. (2013b). Learning in the 21st century: Digital experiences and expectations of tomorrow's teachers. Retrieved from http://www.tomorrow.org/speakup/tomorrowsteachers_report2013.html

Project Tomorrow. (2015). Digital learning 24/7: Understanding technology—Enhanced learning in the lives of today's students. Retrieved from http://www.tomorrow.org/speakup/SU14DigitalLearning24-7_StudentReport.html

SchoolDude. (2013). *The unique challenges facing the IT professional in K–12 education.* Retrieved from https://www.schooldude.com/Portals/0/public%20content/reports%20and%20presentations/technology%20management/sr-2013-it-survey.pdf

State Educational Technology Directors Association (SETDA). (2015). *The guide to implementing digital learning: Technology support.* Retrieved from http://digitallearning.setda.org/tech-support/#!/overview

Sun, J., Heath, M., Byrom, E., Phlegar, J., & Dimock, K. V. (2000). *Planning into practice.* Retrieved from http://www.sedl.org/pubs/tec29/planning-into-practice.pdf

Wheeler, T. (2014). *Closing the Wi-Fi gap in America's schools and libraries* [blog post]. Retrieved from https://www.fcc.gov/blog/closing-wi-fi-gap-america-s-schools-and-libraries

Related Resources

At the time of publication, the following ASCD resources were available (ASCD stock numbers appear in parentheses). For up-to-date information about ASCD resources, go to www.ascd.org.

ASCD Edge®
Exchange ideas and connect with other educators interested in various topics, including on the social networking site ASCD Edge® at http://edge.ascd.org.

Print Products
Digital Learning Strategies: How do I assign and assess 21st century work? (ASCD Arias) by Michael Fisher (#SF114045)
Enhancing the Art and Science of Teaching with Technology by Sonny Magaña and Robert J. Marzano (#313077)
Getting Started with Blended Learning: How do I integrate online and face-to-face instruction? (ASCD Arias) by William Kist (SF115073)
Teaching with Tablets: How do I integrate tablets with effective instruction? (ASCD Arias) by Nancy Frey, Doug Fisher, and Alex Gonzalez (#SF113074)
The Tech-Savvy Administrator: How do I use technology to be a better school leader? (ASCD Arias) by Steven W. Anderson (#SF115015)

ASCD PD Online® Courses
Blended Learning: An Introduction (#PD14OC009M)
Technology in Schools: A Balanced Perspective, 2nd Edition (#PD11OC109M)
These and other online courses are available at www.ascd.org/pdonline.

About the Author

Susan Brooks-Young has been involved in the field of instructional technology since 1979. One of the original technology users in the district where she taught, Susan has continued to explore ways in which technology can be used to facilitate student learning. She has worked as a computer mentor, technology trainer, and technology curriculum specialist.

Prior to establishing her own consulting firm, Susan was a teacher, site administrator, and technology specialist in a county office of education in a career that spanned more than 23 years in both public and private education. Since 1986, she has published articles and software reviews in a variety of education journals. She is also author of a number of books that focus on effective use of technology in schools.

Susan works with educators internationally, focusing on practical technology-based strategies for personal productivity and effective technology implementation in classrooms. Mobile technologies and BYOD programs are areas of particular interest for her. She can be reached at sjbrooksyoung@gmail.com.

WHAT KEEPS YOU UP AT NIGHT?

ASCD Arias begin with a burning question and then provide the answers you need today—in a convenient format you can read in one sitting and immediately put into practice. Available in both print and digital editions.

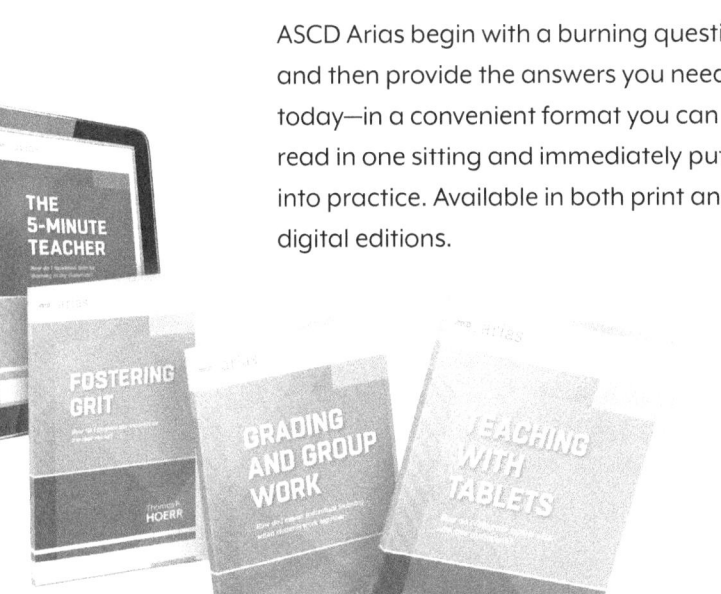

Answers You Need
from Voices You Trust

ASCD | arias

For more information, go to www.ascdarias.org or call (toll-free in the United States and Canada) 800-933-ASCD (2723).

www.ascdarias.org

www.ingramcontent.com/pod-product-compliance
Lightning Source LLC
Chambersburg PA
CBHW070551300426
44113CB00011B/1865